Careers in the Military

Service Careers in the Military

Leanne Currie-McGhee

San Diego, CA

© 2023 ReferencePoint Press, Inc.
Printed in the United States

For more information, contact:
ReferencePoint Press, Inc.
PO Box 27779
San Diego, CA 92198
www.ReferencePointPress.com

ALL RIGHTS RESERVED.
No part of this work covered by the copyright hereon may be reproduced or used in any form or by any means—graphic, electronic, or mechanical, including photocopying, recording, taping, web distribution, or information storage retrieval systems—without the written permission of the publisher.

LIBRARY OF CONGRESS CATALOGING-IN-PUBLICATION DATA

Names: Currie-McGhee, L. K. (Leanne K.), author.
Title: Service careers in the military / by Leanne Currie-McGhee.
Description: San Diego, CA : ReferencePoint Press, Inc., 2023. | Series: Careers in the military | Includes bibliographical references and index.
Identifiers: LCCN 2022010597 (print) | LCCN 2022010598 (ebook) | ISBN 9781678202965 (library binding) | ISBN 9781678202972 (ebook)
Subjects: LCSH: Service industries--Vocational guidance--United States--Juvenile literature. | Veterans--Employment--United States--Juvenile literature. | United States--Armed Forces--Vocational guidance--Juvenile literature. | United States--Armed Forces--Job descriptions--Juvenile literature.
Classification: LCC UB147 .C873 2023 (print) | LCC UB147 (ebook) | DDC 355.002373--dc23/eng/20220325
LC record available at https://lccn.loc.gov/2022010597
LC ebook record available at https://lccn.loc.gov/2022010598

Contents

Introduction: Need for Service	4
Logistics Specialist	8
Nutrition Care Specialist	16
Retail Services Specialist	23
Food Service Specialist	30
Finance Officer	37
Public Affairs Officer	43
Source Notes	50
Interview with a US Navy Public Affairs Officer	53
Other Service Careers in the Military	56
Find Out More	57
Index	59
Picture Credits	63
About the Author	64

Introduction: Need for Service

It is early morning and still dark out on the ocean, but an industrial-sized kitchen is full of people in chef's hats and US Navy uniforms. They are slicing meat, grilling vegetables, and baking bread. Aboard the aircraft carrier USS *Harry S. Truman*, these culinary specialists are preparing to feed 5,500 sailors. In one day, a kitchen crew of 114 sailors makes approximately 17,300 meals. The entire time they are at sea, sailors on the USS *Harry S. Truman* depend on these sailors for all of their meals.

At the same time, logistics personnel are planning and arranging for fuel, food, and supplies to be delivered to all ships, including the USS *Harry S. Truman*. These personnel are determining how much fuel, food, and supplies each ship has; how long these will last; and when and how to replenish the ships. "Every U.S. Navy Sailor out there knows they can rely on our logistics team," says Captain Chuck Dwy, an assistant chief of staff for logistics. "No matter how rough is the sailing or in spite of a global pandemic, we will get you what you need—the food, the fuel, the ordnance, the parts—to stay in the fight."[1]

This is just a sampling of men and women who work in service careers in the military. Service careers are those careers in which people or companies do not make a product but instead provide a needed service to others. In all branches of the military, whether on a ship or a base or in the field, services like food preparation, medical services, and supply planning are required to keep operations running. Other services include legal services, financial services such as payroll and accounting, and psychological counseling. Without such services and those who provide them, the military could not operate.

Behind the Scenes

When people think of jobs in the military, their first thought is probably of pilots flying helicopters and jets, soldiers on reconnaissance missions, and sailors steering ships, aircraft carriers, and submarines. These jobs directly accomplish the main mission of protecting and defending the United States. However, for any of these jobs to occur, support services are needed. These services keep the soldiers and sailors motivated and ready to serve.

Some of these services are focused on keeping military members healthy. As an example, on navy ships, the medical staff treats all manner of sickness and injury. On the USS *America*, the navy's newest ship, there are two operating rooms, a three-bed ICU, and a twenty-three-bed area for sailors to recover while the ship is under way. There are also three dental wards on the ship. Navy doctors, dentists, corpsmen (enlisted personnel trained in first aid and basic medical services), and nurses are assigned to the ship. Medical staff are also assigned to bases, where they provide medical care to active-duty military and their family members. "We serve in hospitals, on ships above and below the surface, and follow our Marine brothers and sisters in every operational area,"[2] says Todd Wende, a former master chief hospital corpsman.

Military support services also include personnel who provide mental health assistance to individuals and families. Military counselors and psychologists work with active-duty personnel, veterans, and family members on issues ranging from depression and anxiety to substance abuse to post-traumatic stress disorder. Navy psychologist James McClelland served on a mission in 2021 that provided health services to military personnel and their families in Barlow, Kentucky. "Folks will come in with a lot of different issues," says McClelland. "Depression, anxiety. They may come in with family or financial stress and may really need to just talk about it. Additionally, I help connect them with resources in the community, such as mental health providers."[3]

Sample Military Pay Scales, 2022

Basic pay for military personnel, whether enlisted or officers, is based on years of service and rank. The person's rank usually corresponds with his or her pay grade. Individuals with more years of service and higher rank achieve higher pay grades. As in the civilian world, basic pay is subject to taxes. Some military personnel supplement their income with allowances for housing, clothing, and other needs. Special and incentive pays, such as for hardship duty, can also increase income.

A Sample of Monthly Active-Duty Enlisted Pay Scale for 2022 (all branches)

Sample Pay Grades	Years of Service						
	<2	2	3	4	6	8	10
E-2	$2,054.72	$2,054.72	$2,054.72	$2,054.72	$2,054.72	$2,054.72	$2,054.72
E-3	$2,160.71	$2,296.58	$2,435.84	$2,435.84	$2,435.84	$2,435.84	$2,435.84
E-4	$2,393.32	$2,515.94	$2,652.12	$2,786.76	$2,905.38	$2,905.38	$2,905.38
E-5	$2,610.22	$2,786.15	$2,920.79	$3,058.51	$3,273.25	$3,497.55	$3,682.10
E-6	$2,849.31	$3,135.53	$3,274.18	$3,408.51	$3,548.70	$3,864.19	$3,987.74
E-7	$3,294.21	$3,595.53	$3,733.56	$3,915.33	$4,057.99	$4,302.62	$4,440.65

A Sample of Monthly Active-Duty Officer Pay Scale for 2022 (all branches)

Sample Pay Grades	Years of Service						
	<2	2	3	4	6	8	10
O-1	$3,477.22	$3,619.56	$4,375.64	$4,375.64	$4,375.64	$4,375.64	$4,375.64
O-2	$4,006.53	$4,562.65	$5,254.95	$5,432.73	$5,544.26	$5,544.26	$5,544.26
O-3	$4,636.60	$5,255.88	$5,672.43	$6,185.42	$6,482.12	$6,807.16	$7,017.29
O-4	$5,273.75	$6,104.39	$6,512.31	$6,602.58	$6,980.62	$7,386.39	$7,891.67
O-5	$6,112.09	$6,885.42	$7,361.74	$7,451.40	$7,749.33	$7,926.80	$8,318.08
O-6	$7,331.86	$8,054.66	$8,583.36	$8,583.36	$8,616.32	$8,985.43	$9,034.42

E = Enlisted   **O** = Officer

Source: Brittany Crocker, "2022 Military Pay Charts," The Military Wallet, December 29, 2021. https://themilitarywallet.com.

After the Military

Those who work in military service careers often have a fairly easy transition to civilian life. Many of these military jobs have obvious civilian equivalents. The skills required for cooks, accountants, warehouse managers, therapists, and other jobs in the service industries are essentially the same whether those jobs are military or civilian.

Those who serve as retail services specialists may be able to obtain retail management jobs. Officers who worked in finance may be able to transition to careers in banks or in corporate finance. Food service specialists find that restaurants and dining institutions need food preparation specialists, bakers, cooks, and dining managers. And because of the level of responsibility young adults are given in the military, their experience can lead to careers with paths toward higher-paying jobs.

Serving in the military has its challenges, such as moving every few years, deploying away from home for several months at a time, and being legally obligated to serve for a certain number of years. Plus, those who serve during wartime are likely to encounter additional risks and responsibilities. However, the training, responsibility, chance to see the world, and service to country provide reasons why those thinking about a career in services should consider the military.

Logistics Specialist

What Does a Logistics Specialist Do?

Whether purchasing and stocking supplies in a base warehouse, ordering new parts for an army unit's vehicles, or arranging for flight gear for aircraft, logistics specialists ensure military personnel have the supplies they need to do their jobs. "Everybody has to come to supply to get the items that they need," explains a navy logistics specialist who works on an aircraft carrier. "We order generators, transmitters, a lot of electronics . . . tailhooks, [and] tires to keep the aircraft moving."[4]

These enlisted personnel are the ones responsible for purchasing, reviewing the orders when received, and arranging for transport of material such as wood or wiring, equipment, and supplies. For example, if assigned to a parts warehouse, they perform duties such as checking inventory of current supplies, ordering new supplies as parts run low, and unloading and stocking the supplies. The logistics specialist might also be responsible for tracking money coming in and going out during the ordering process. It is a job with a whole range of responsibilities, as navy logistics specialist Nathan Grant explains. "One of the greatest things about

A Few Facts

Minimum Educational Requirements
High school diploma or general equivalency diploma

Personal Qualities
Organized, detailed, personable

Working Conditions
In offices on bases, at a warehouse, on ships or submarines

Salary
Depends on pay grade and years of service

Future Job Outlook
Approximately 29,000 in the military and increasing

being in the Military is that you can have a job, like being a logistics specialist, and you can do inventory for a year. You can go off and do accounting. You can be a purchasing manager. It really gives you a broad spectrum of opportunity within one job."[5]

The job has different titles in different branches of the military and, sometimes, slightly different roles. In the marines, logistics/embarkation specialists oversee the loading process for ships and aircraft in addition to monitoring and ordering supplies and equipment. In the army, logistics specialists are more specialized. Some might work with medical units. Others might work with automated systems, where they do all of the ordering and planning of materials with specific software. In the air force, the job title is logistics plans specialist. Regardless of which branch they work in, all logistics specialists are focused on making sure military personnel have the supplies and equipment they need to do their jobs.

A Typical Workday

Before enlisting in the navy, Angelica Gonzalez Flores worked as a medical social worker in Los Angeles. Now she is a navy logistics specialist aboard the hospital ship USNS *Mercy*. The *Mercy* is a one-thousand-bed hospital ship that provides acute surgical and medical facilities for military-specific missions as well as disaster relief and humanitarian assistance worldwide. In 2020, as hospitals nationwide were overrun with COVID-19 patients, the *Mercy* was tasked with helping relieve the burden on hospitals. It was deployed to San Diego, where it provided medical services to non-COVID-19 patients. With resources at civilian hospitals devoted almost entirely to the COVID-19 emergency, patients with other health issues were encountering difficulty getting the medical care they needed. On board the *Mercy*, Flores worked in the ship's store, making sure that medical personnel had bandages, masks, syringes, blood-pressure cuffs, and other supplies and equipment needed for treating patients with various health conditions. She also tracked inventory to make sure the ship did not run out of those necessities.

The typical day of a logistics specialist who works on a military base is a little different from the typical day aboard the *Mercy* or other ships. Navy logistics specialist Grant is responsible for ordering aviation supplies for all different command groups on his base. He works closely with each command group to determine the type and quantity of supplies and equipment they will need. He handles contracts for those purchases and records and tracks all orders and purchases. His job also entails making sure that command groups do not exceed their budgets. Grant says:

> On a typical day, I obligate money to different accounts, contact vendors, set up contracts. We manage aviation parts for all the different commands within the flight line, the helicopters, and the planes. We also manage official travel and then anything needed to run regular facility services.
>
> We normally have about 20 commands we support directly, and then we do an assessment prior to the start of the fiscal year and ask them what their requirements are. The easiest way to do it is look at the prior history, and when they give us their projections, we will go back and say, "Oh, well, last fiscal year you spent this amount of money." And then we'll look at what they're giving us. We're using taxpayer dollars, so we're trying to be cost-efficient and effective with the money.[6]

Other logistics specialists work in on-base warehouses. Anna Brandt is a logistics specialist and army reservist who works at the Sioux City Army Air Base. "What we do is warehouse operations. If [units on the base] need a part . . . we . . . order it, we supply it,"[7] Brandt explains. She maintains an inventory of aircraft parts that are ordered, received, and distributed for base operations. Additionally, her tasks may include unloading, unpacking,

Like Mother, like Daughter

Two navy logistics specialists were especially excited to serve on the same ship for four weeks. This is because they share a special bond as mother and daughter. Master chief logistics specialist Tonya McCray and her daughter, logistics specialist seaman Racquel McCray, spent four weeks together on the USS *Gerald R. Ford*, where Racquel was sent for training. Racquel grew up seeing both of her parents serve in the navy and ultimately wanted to follow in their footsteps, either in information technology like her father or as a logistics specialist like her mother. "I chose to join because I watched my parents for my entire life," Racquel explains. "They both served, so watching them every day go to work made me actually want to follow in their footsteps, with how successful they were and what they were able to provide for my sister and I." For Tonya, this provided an opportunity to show her daughter what she can aspire to if she sticks with the job. "I was able to share what I did with my daughter every day," Tonya explains. "She saw what respect that someone of my pay grade gets on a day to day [and] how they look up to me."

Quoted in Joelle Goldstein, "Meet the 'Proud' Mother and Daughter Who Got to Serve in the Navy on the Same Ship Together," Yahoo!, August 10, 2021. www.yahoo.com.

and shelving supplies when inventory arrives, checking to see whether the inventory matches what was ordered, and ensuring all is tracked on their computer system. Like all logistics specialists, her main mission is to ensure that the military members have the supplies needed to fulfill their duties.

Education and Training

Logistics specialists must obtain a high school diploma or general equivalency diploma before enlisting in the military. Once enlisted, all go through basic training, where they learn the basics of being in the military, from marching and getting physically fit to military history and weapons training, at their respective branch's training base. Following basic training, logistics specialists receive specialized training for their job. Logistics specialists in the navy are primarily trained at the Naval Technical Training Center in Meridian, Mississippi, for nine weeks after basic training. In the army, these specialists attend twelve weeks of training at Fort Lee in Virginia. In the air force, these personnel are called logistics plans specialists. They attend school at Lackland Air Force Base in Texas for approximately thirty days. In all branches, logistics specialists learn how to contract with vendors, distribute supplies and equipment, monitor inventory, and maintain financial records.

A US Navy logistics specialist guides a forklift on the flight deck of an aircraft carrier. Logistics specialists handle purchasing, review items received, arrange for transport, and unload and stock incoming supplies.

Skills and Personality

Logistics specialists interact with other people on a daily basis. They work directly with personnel who request supplies and equipment and with vendors or suppliers. For this reason, being personable and customer oriented are traits that will assist them in doing their job.

In a job that requires working with inventory, tracking and logging needs and distributions, and maintaining financial records, organizational skills and attention to detail are important characteristics. "I keep track of parts, I receive parts. They [units] come to me to order them, especially for vehicles that break down. They'll bring me a work order, and I'll order that part,"[8] an army logistics specialist explains.

Working Conditions

Logistic specialists can be assigned to offices in shore bases or on ships or submarines. They also may work in warehouses or a ship or base store. On shore duty, their hours are typically during the day. If deployed overseas or on a ship, they may work different shifts and either day or night hours.

Depending on their specific duty, some may spend much of the day on a computer, tracking and ordering parts. Others may be in a ship or base store, interacting with military personnel as they provide the personnel with supplies from the store, stocking the shelves, or updating inventory on a computer. Those in warehouses may be involved with actually loading and unloading supplies and equipment as these arrive or are distributed.

Some logistics specialists work with hazardous materials. Those who do this type of work receive special training on hazardous material control and management. This training allows them to organize and maintain databases, files, and reports related to the hazardous material program and follow safety procedures when actually dealing with any of the materials.

The job of a logistics specialist can involve deployments to bases overseas or on ships sailing the world's oceans. "Travelling

COVID-19 Challenges for a Logistics Specialist

"Materiel Management—Supply department is critical in fighting the spread of COVID 19. Specifically, we are overall responsible for ordering and tracking PPE [personal protective equipment] that protects the entire hospital staff. Acquiring supplies, in general, has been a hurdle worldwide. I am lucky to have an amazing team in Material Management that works around the clock to ensure that our supplies are always up to date to support all the clinics. We have to consider who needs what and how we can substitute if we don't have the requested item. It has been challenging to find vendors for supplies that cannot be filled by our prime vendor. There are no breaks in searching for PPE supplies."

—Brenda Kedar Ike, navy logistics specialist assigned to Navy Medicine Readiness and Training Command in Bremerton, Washington

Quoted in Douglas Stutz, "I Am Navy Medicine Helping to Stop the Spread of COVID-19: Logistics Specialist 1st Class (SW/AW/IW) Brenda Kedar Ike, NMRTC Bremerton," DVIDS, July 23, 2020. www.dvidshub.net.

is the coolest part of my job. I've travelled to Italy, Spain, South Africa, Bahrain, Dubai. My last deployment I've been to Jerusalem. I've been a few places,"[9] says a navy logistics specialist.

Opportunities for Advancement

Enlisted personnel have the opportunity to advance from the entry level, E-1, to the highest enlisted level, E-9. As one advances, the requirements to achieve the next promotion are more challenging. Achieving this requires receiving positive fitness reports, obtaining higher-level education such as an associate's or bachelor's degree, and taking on additional military duties, such as leading physical fitness training. As one advances, examples of different types of jobs are warehouse supervisor, post office manager, and department supply chief.

Employment Prospects in the Civilian World

A logistics specialist has skills that are transferrable to positions involving inventory, supplies, and financial tracking. Many civilian companies hire veteran logistics specialists to work in positions such as inventory clerk, warehouse manager, purchase order clerk, and budget analyst. Major companies like Amazon specifically hire for logistics specialist positions and encourage veterans to apply.

Nutrition Care Specialist

What Does a Nutrition Care Specialist Do?

"Everyone has to eat, and what we eat and how much we eat can have a real impact on our physical and mental wellbeing,"[10] says army officer Stephanie Gasper. Gasper is program director of the nutrition and diet therapy training for army nutrition care specialists and air force diet therapists at Joint Base San Antonio–Fort Sam Houston. The nutrition care specialists (enlisted) who pass through the program work to ensure that military personnel are eating healthful, well-balanced diets. They also work with personnel who have chronic conditions or are recovering from illness or injury. Rapid recovery, disease management, and the promotion of overall health and wellness are central to a healthy, well-functioning military.

In military hospitals and clinics, these specialists might create meal plans for patients who have chronic conditions such as high blood pressure or diabetes. Diabetes, for instance, requires careful monitoring of food intake and blood sugar levels. Additionally, people with

A Few Facts

Minimum Educational Requirements
High school diploma or general equivalency diploma

Personal Qualities
Good communicator, personable, detail oriented

Working Conditions
In hospitals, wellness centers, clinics

Salary
Depends on pay grade and years of service

Future Job Outlook
Fewer than 1,000 in military; competitive field

diabetes are typically restricted from eating certain types of foods. Nutrition care specialists also help coordinate diet education and connect personnel with other specialists who can help them with their dietary needs. Additionally, nutrition care specialists may be involved with the actual preparation of meals for their patients, working in the kitchen with food service specialists to ensure that the meals made meet patients' nutrition requirements.

Brandon Sims is an air force diet therapist (which is that branch's title for a nutrition care specialist). He explains, "Every other day we have classes for [patients]. . . . We also have one on one appointments with our dieticians if people aren't understanding what we are teaching them in class."[11] Specialists like Sims work with registered dietitians to develop individualized diet plans and specially balanced menus for patients.

Nutrition care specialists also help teach all military members wellness, nutrition, health, and good food choices. Military members can consult with them at base wellness centers to help them plan a diet that will improve their health. Also, they may be consulted to help optimize the fitness levels of soldiers or those going into the field. According to Gasper,

> Military nutrition technicians [nutrition care specialists] can work in food service operations and medical field feeding, sports nutrition to optimize performance and support the warfighter, nutrition for general health and wellness or disease prevention or perform patient care through medical nutrition therapy for diseases or other conditions in both a hospital and inpatient or ambulatory settings.[12]

A Typical Workday

Anthony Profit is a nutrition care specialist in the army, assigned to the 212th Combat Support Hospital in Germany. His workdays (or worknights) are filled with different tasks, ranging from conducting nutrition assessments to actually preparing meals for

> **Fueling the Body**
>
> "My job is understanding how to fuel for performance, supplementation and assisting patients with nutrient timing. The best way to maintain nutritional fitness is to understand what your body needs."
>
> —Christopher Gadson, army nutrition care specialist at Bayne-Jones Army Community Hospital in Louisiana
>
> Quoted in Angie Thorne, "Total Force Fitness: Fuel Body, Optimize Performance," DVIDS, January 8, 2021. www.dvidshub.net.

patients. Profit explains that part of his job is also working in the kitchen to create the meals he has planned for the patients. As a senior enlisted member, he leads the other enlisted personnel in the nutrition care division, ensuring that they are meeting the patients' nutrition needs.

Although nutrition care specialists play a big part in helping patients manage chronic conditions and recover from illness and injury, many military personnel do not know about this group of specialists or what they do. "I do wish that more service members were aware of what we do nutrition-wise," says Profit. "They don't see us when we do a nutrition assessment with patients or when we are putting together patient trays, interviewing patients periodically, and ensuring that they are receiving the right nutrition therapy for whatever diagnosis they have when they are in a hospital."[13]

What Profit and other nutrition care specialists like about the job, particularly when based at a hospital, is interacting with patients and seeing their progress. "I like working with patients and seeing them improve gradually," says nutrition care specialist Shane Rose. "It gives you a sense of satisfaction and pride that you can help this person recuperate and get back to their best."[14]

At Landstuhl Regional Medical Center in Germany, air force nutrition care specialist Nataliya Hampton says that her workdays include making sure that patients get the meals that are designed for their specific medical needs. She does this by working with a team of medical and culinary professionals. Hampton explains:

> If you're a diabetic, you can't just eat whatever you want. The physician prescribes a diet order; we provide them an actual medical nutrition therapy–based diet. It's amazing to see a team come together from the culinary side of the house and the inpatient feeding side of the house to ensure the operation happens and the safety of the patients. We [spin] up resources to ensure quality and safe nutrition [are] being provided.[15]

The medical center sees a variety of patients, including military personnel based nearby, medical evacuees from combat zones, and other people who work with the military. These patients come to the medical center from Europe, Africa, and the Middle East. Each person's nutrition needs are assessed by

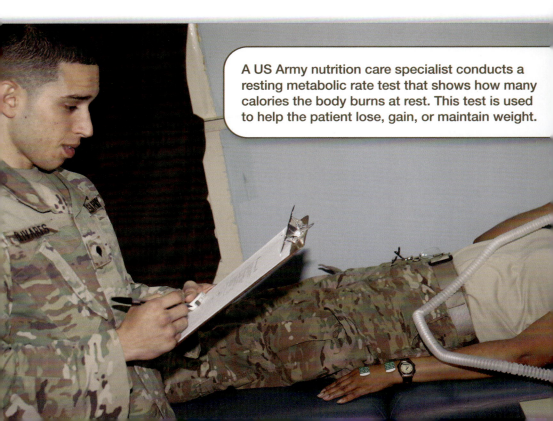

A US Army nutrition care specialist conducts a resting metabolic rate test that shows how many calories the body burns at rest. This test is used to help the patient lose, gain, or maintain weight.

nutrition care specialists like Hampton, under the supervision of a dietitian. These specialists take pride in the fact that they are an integral part of maintaining the health of the military.

Education and Training

Those who wish to pursue this career path need to have a high school diploma or general equivalency diploma. Candidates need a high score in the operators and food segment of the Armed Services Vocational Aptitude Battery, a test all enlisted recruits must take. This test scores candidates in different areas, helping the military match them to an appropriate job. In each military branch, after completing basic training, nutrition care specialists receive additional training in their field. Army and air force nutrition care specialists attend eight weeks of advanced training at Fort Sam Houston in Texas. They are taught how to complete nutrition screenings of patients, implement nutrition therapy for medical conditions, plan menus, operate and clean food service equipment, and procure, store, and prepare dietetic foods and supplies. Additionally, they learn to prepare a specific meal based on a nutrition plan that has been designed by a dietitian for a patient. Nutrition care specialists are trained in cooking techniques (because sometimes they do cook for their patients or at least work with food specialists who do the cooking). They also study how to determine what foods and meals will benefit patients with various health needs.

Skills and Personality

Nutrition care specialists must have the ability to learn to cook and prepare food. They also must have an interest in and ability to understand the science behind nutrition in order to assess individual nutritional needs. Tied to this, nutrition care specialists must be detail oriented to ensure that they properly prepare the meals for their patients. They must measure and weigh food to get the exact amounts and must ensure that they are following the diet restrictions, such as low sugar or low salt, for their patients.

Focusing on Child Health

The Weed Army Community Hospital Nutrition Care Division at Fort Irwin in California focuses on not just military members but also the health of their families. Nutrition care specialists do outreach work to encourage a healthy lifestyle for military members' children, specifically focusing on reducing childhood obesity. As part of this effort, they promote the base's after-school and summer programs to keep kids active. They also see pediatric patients at the hospital and help the children's parents develop a healthy diet for them. "You can see the emotional distress in a child that is overweight, but when they take actions to get the weight off there's an inherent self-esteem boost because they feel better mentally, they look better and they are included in activities," says army nutrition care specialist Michael L. Polmanteer.

Quoted in Jalonda Garrison, "WACH Nutrition Care Division Getting Out and Active with Fort Irwin Youth," DVIDS, August 27, 2018. www.dvidshub.net.

Because this job requires much interaction with patients, nutrition care specialists should be personable and encouraging. They need to be able to listen to their patients and effectively communicate to the patients their personalized nutrition plans and the reasons for the plans. Additionally, they often work as a team and with others in the medical field, so being able to get along well with others is necessary.

Working Conditions

Nutrition care specialists can be based at military locations throughout the world, typically in hospitals or field hospitals, clinics, and wellness centers. During their shifts, they may be with patients taking assessments and explaining the diets or in the kitchen preparing food. They also may use food preparation equipment and appliances such as blenders, steamers, ovens, and ranges as a part of food preparation.

Generally, they work with a team of other nutrition care specialists and under the supervision of a dietitian. They may also interact with doctors, nurses, and other medical professionals who are overseeing the patients. Additionally, they may work with culinary specialists who do general food preparation, providing input on the special preparation of meals for patients.

Opportunities for Advancement

As with all enlisted military personnel, nutrition care specialists are able to work toward advancement from entry level, E-1, to the highest enlisted level, E-9. To achieve higher ranks, they must work toward excelling at their job and showing dedication to the military. High positive fitness reports, involvement in their unit's activities, and obtaining higher-level education, such as an associate's or bachelor's degree, are ways that help them advance. As these specialists advance, examples of different types of leadership positions include nutrition care division supervisor, lead nutrition care specialist, and nutrition care instructor.

Employment Prospects in the Civilian World

Nutrition care specialists can transfer their skills to civilian jobs involving food and nutrition. Some are able to pursue work as a line cook or sous-chef in a restaurant because of their food preparation background. With a nutrition background, a veteran can pursue a career as a dietetic technician at wellness centers and medical clinics or a dietary cook in a hospital. Additionally, veteran nutrition care specialists have a solid background to pursue further education and certification to become a registered dietitian.

Retail Services Specialist

What Does a Retail Services Specialist Do?

Keeping up morale is important because it contributes to military personnel giving their full effort to their job. Retail services specialists help achieve this by providing military members with items and services that help them maintain or improve their quality of life. From providing coffee on a ship to disc golf at a recreation center on a base, retail services specialists interact directly with other military members in ways that boost their morale.

Retail services specialists manage and oversee shipboard and base retail and service activities, including the stores, vending machines, coffee kiosks (on aircraft carriers), barber shops, and laundry operations. They may work in recreation or fitness centers on base that provide recreation services to military members and their families. They are the ones who directly interact with the personnel, much like a person working retail in a store or restaurant.

Their work often includes checking out items for customers, taking payment, and helping

A Few Facts

Minimum Educational Requirements
High school diploma or general equivalency diploma

Personal Qualities
Outgoing, organized, versatile

Working Conditions
In base or ship stores, barber shops, laundry ships, recreation centers

Salary
Depends on pay grade and years of service

Future Job Outlook
Competitive field

customers with finding what they need. They use handheld scanners to track inventories, check and process orders and receipts, and complete inventory adjustments. They may also order, receive, check, store, and keep inventory for these stores. Their job also entails keeping track of items entering or leaving storerooms and reporting damaged or spoiled goods.

A Typical Workday

The most popular places on a ship are the vending machines, ship's store, and coffee shop. And these are the places where you will find retail services specialists like Liza Pangborn. Pangborn is stationed on the USS *Nimitz*, an aircraft carrier with over three thousand sailors. Her day includes ensuring these sailors have access to items that help them feel more comfortable while at sea. She often tends the coffee shop—brewing coffee and serving it to the sailors—or works in the ship's store selling items from energy drinks or games to toiletries. Pangborn says:

> My job is to make sure the sailors on *Nimitz* have the comforts of home. To some, we just make coffee or operate the ship's store, but to the sailors we serve with we offer normalcy out at sea. That means a lot to me because my shipmates sacrifice so much and we're able to give them a morale boost. We see over 500 Sailors a day underway. They're sleepy while they put their order in and the moment they have that first sip of coffee you can see the change in mood.[16]

On bases, retail services specialists may serve in base exchanges (retail stores for service members) or recreational facilities. Joshnino Supapo has done both. Supapo works as a Marine Corps Community Services (MCCS) specialist—equivalent to a retail services specialist in the navy. He manages the exchange at a marine base in Quantico, Virginia. Before that he was an assistant manager at a base golf course. As a store manager,

Transition to the Civilian World

Retail services specialists in the navy had a different job title up until 2019. Prior to that, their rating (job title) was ship serviceman. In the civilian world, not a lot of people knew what that meant. So the title was changed to smooth the transition to civilian work. Now when veterans' résumés are reviewed by potential employers or they are asked about prior jobs in job interviews, they can use the title retail services specialist. Potential employers understand that the job candidate's duties were similar to a person working in a store as a clerk or store manager. "By putting retail services specialist on a resume or job application, a person is going to know the line of work I was in," says retail services specialist Tawian Buford. "They're going to see that not only was I in the Navy, but what [I] did specifically for it and the expertise I would be bringing to their business."

Quoted in Kyle Carlstrom, "Retail Services Specialists Embrace New Rate Name," DVIDS, October 19, 2019. www.dvidshub.net.

Supapo handles staff scheduling, helps customers, and makes sure that the exchange is stocked with the items people need and want. Sometimes the job takes these specialists to conflict zones (or forward environments), where they also provide these services. "Being an MCCS Marine is a unique experience because we are given opportunities to provide services to Marines in the rear conducting the humbling tasks such as cashiering, stocking and cleaning, which we do in a forward environment," Supapo says. "We pride ourselves in knowing that we as Marines will take care of our brothers and sisters, no matter where."[17]

Supapo and others who work in these jobs often have busy days and a lot of customer interaction. One retail services specialist, who is stationed on a ship and typically assigned to the ship store or cutting hair in the barbershop, explains what his day is like. "A day usually starts off by coming to the store and mak[ing] sure all the barcodes [on the store items] are up, mak[ing] sure everything scans, and everything is in its right place for when the store opens and is presentable to the crew [sailors on the ship]. Everyone comes in and starts shopping and we provide the customer service, and everyone asks [us] if they need help with anything."[18] For retail services specialists, helping military personnel and raising their morale is the main goal. This is what provides them much job satisfaction at sea, in the field, or on the base.

Education and Training

All in this job must first obtain a high school diploma or general equivalency diploma. After being accepted into the military, candidates attend basic training. For those in the navy, this job requires a five-year service obligation after leaving for boot camp. Following boot camp, in the navy, retail services specialists attend advanced training school for five weeks in Meridian, Mississippi. At school they learn what tasks are required for being a ship's store retail operator, a barber, and a shipboard laundry operator. (These

A US Navy retail services specialist cuts hair in the barber shop on board an aircraft carrier. Retail services specialists work in a variety of jobs, all of which help improve quality of life for active-duty personnel.

jobs might sound simple, but things can get pretty involved when you are providing these services for hundreds or thousands of people every single day.) Marine corps members in this job are referred to as community service specialists. The marines receive on-the-job training, and only those who have served in other job areas first may apply to be a community service specialist. The on-the-job training is six months long, during which they are assigned to a location and learn the job under the leadership of another higher-ranking member. During on-the-job training, marines learn about finance, human resources, loss prevention for inventory, warehouse operations, store operations, and store manager operations.

Skills and Personality

Because their days are filled with customer interactions, retail services specialists need to be personable and able to get along with people. They must be able to communicate and listen effectively so they can provide their customers with the specific items or services they need or want. As in civilian life, a military barber needs to listen to his or her customer to know what haircut to give the client.

Being organized and detailed is necessary since the job often entails organizing supplies in stores, taking inventory of items in the storeroom, and tracking what is sold and needs restocking. Additionally, retail services specialists must have reasonably good math skills because they deal with all sorts of sales transactions.

Retail services specialists need to be versatile because their duties can change depending on what is needed. "We do a little bit of everything—the ship's store, the barbershop, vending machines, and then, laundry. We also do staterooms [officer bedrooms on ships], maintaining the cleanliness of staterooms, and making sure the officers have sheets . . . everything I do, I make sure I do it right,"[19] explains retail services specialist Evan Trujillo.

Working Conditions

Retail services specialists who are assigned to ships live in dorm-like rooms with other enlisted sailors. They live on the ship and can work out at the gym, hang out in the recreational areas, and eat in the dining hall when not working. Retail services specialists on ships may be on call day or night, depending on their duty hours. On ships, they may be assigned to the ship's store, laundry services, barber shop, or coffee shops.

On base, retail services specialists may work in the base stores, or if in the marines, they may be assigned to recreational services like a golf course. Their day typically includes much customer interaction, since they are the ones who provide the actual items or services to the personnel. They spend much of their time on their feet, cutting hair, assisting customers, stocking store items, cleaning clothes in the laundry, or tracking inventory. Additionally, depending on their duties, they may work with equipment such as laundry machines for washing, drying, and pressing; coffee and espresso machines; and cash registers.

Opportunities for Advancement

Retail services specialists can advance from entry level, E-1, to the highest enlisted level, E-9. In order to do this, a retail services specialist needs to obtain high fitness reports, take on collateral military activities outside of his or her job, and pass tests that are required for higher-level ranks. Obtaining an associate's or bachelor's degree will also help retail specialists advance. Those who achieve the level of E-6 may be able to apply for commissioning as a limited duty officer, which is a higher level than all enlisted ranks. They can advance to positions such as assistant store manager, store manager, laundry supervisor, and retail services specialist instructor.

Becoming a Barber

"It's a cool feeling to transform somebody's entire look just by cutting their hair. It's like art, like trying to make a masterpiece. When I'm in the barbershop it feels like I'm by myself, doing my own thing, and I'm able to create something and make someone look decent at the same time."

—Devonta Allen, navy retail services specialist

Quoted in Ford Williams, "More than Just a Haircut," DVIDS, January 4, 2017. www.dvidshub.net.

Employment Prospects in the Civilian World

Careers in retail, ranging from clerks to store manager positions, are among the job possibilities when retail services specialists leave the military and reenter the civilian world. Experience gained in the military puts them at an advantage for jobs at coffee shops, barbershops, laundry and dry-cleaning services, recreational facilities, and other retail service–based businesses.

Food Service Specialist

What Does a Food Service Specialist Do?

A young person who has a knack for chopping, mixing, baking, frying, braising, simmering, or steaming—or who wants to acquire those culinary skills—might want to consider a job in the military as a food service specialist. Army and marine corps food service specialists (known as culinary specialists in the coast guard and navy and service specialists in the air force) help plan and make the meals for the forces—wherever they are. That can mean aboard an aircraft carrier, on a submarine, in the field, or at a dining hall on base. Navy food service specialists may also serve at Camp David, the retreat for the president and the president's family, and at the White House.

Food service specialists are like cooks anywhere, except that they do a little bit of everything. This includes food preparation, cooking and baking, serving, and cleanup. Their days typically begin early in the morning as they start preparation for breakfast. Food service specialists gather all of the needed ingredients and then start chopping,

A Few Facts

Minimum Educational Requirements
High school diploma or general equivalency diploma

Personal Qualities
Creative, calm, focused

Working Conditions
In base, ship, or submarine kitchens

Salary
Depends on pay grade and years of service

Future Job Outlook
Growing field in the military

mixing, blending, cooking, and baking. As personnel begin filing in for breakfast, these specialists serve up plates of food. Once breakfast is over, the food service personnel clean up the dining hall and kitchen and begin the process of preparing lunch. This happens again at dinnertime and then repeats the next day and the next and the next.

These specialists, who are in the enlisted ranks, also have the responsibility of maintaining clean, safe, and sanitary food service spaces in the kitchen and dining areas. They oversee the cleanliness and organization of storerooms and refrigerated spaces. As they gain more responsibility, food service specialists might be involved in ordering the food needed for meals. They might also help plan the rotation of meals to be served. Their job culminates in feeding the forces, keeping them healthy and satisfied.

A Typical Workday

Air force food service specialist Linsun Jackson, stationed in South Korea, begins her workdays in the dark hours of morning and continues until breakfast, lunch, and dinner are served. "We would start prepping early in the morning for breakfast, serve it, and progressively cook the meals as we need it," says Jackson. "Once breakfast ends, the process starts again for lunch and dinner."[20] After the meals are prepared, Jackson helps serve the meals, and then keeps the kitchen and dining area clean and presentable. She says she receives great satisfaction from seeing air force personnel eat and enjoy their meals.

Like Jackson, Patricia Miller—lead food service specialist on the navy destroyer USS *Preble*—starts work early. She starts preparations for breakfast by assigning jobs to the other specialists. On a ship like the *Preble*, the kitchen is smaller than on a base dining hall, so Miller and the other food service specialists must use every inch of kitchen space to create meals for hundreds each day. Throughout the day, Miller and her food service specialists work side by side chopping vegetables, preparing pizza toppings,

Working in a fully-stocked base kitchen, a US Marine Corps food service specialist removes Italian sausage and peppers from an oven. Food service specialists do food preparation, cooking and baking, serving, and clean-up.

and frying chicken wings. They do this while listening to music. Once the meal is ready, they open the roll-up doors to the dining area and begin serving the line of crew members.

Miller says she and other food service specialists get a lot of satisfaction from seeing crew members enjoy their meals. Food service specialists, she says, "are definitely the heart of the crew, because when you see someone come through the line and they're happy about the product they tasted, it looks good and it literally made their day, it makes our day."[21]

What also makes Miller's job fulfilling, she says, is seeing the positive effects that good food and good cooking have on service members. She says:

> We play a highly significant role in that people having a bad day can come through the line and it will be something their mom used to make all the time. They'll eat it and it will take them back to a place of peace and they will be happy

for the rest of the day. And we'll actually hear about it all day. That one product that just literally turned their whole day around. It's a good feeling. It's a real good feeling.[22]

Food service specialists who are stationed on aircraft carriers might have somewhat larger kitchens than those on destroyers, but they also have the added challenge of feeding thousands of crew members three times a day, every day. In a single day, for example, food service specialists aboard an aircraft carrier will use approximately 1,600 pounds (726 kg) of chicken, 160 gallons (727 L) of milk, 30 cases of cereal, and 350 pounds (159 kg) of lettuce.

When units are in the field, food service specialists often work in temporary kitchens. When army food service specialist Darren Shastid was stationed in the California desert during training

Serving Meals in a Pandemic

Prepping, cooking, and serving meals for large groups of people is always challenging, and the COVID-19 pandemic has made it even more so. However, the military has adapted in order to feed the troops while reducing the risk of COVID-19 transmission. As an example, in 2020 army food service specialists in Hawaii followed COVID-19 protocols while making all the meals for an infantry division that was undergoing training. In addition to standard hygiene practices used in any commercial kitchen, such as gloves and hand washing, the kitchen staff had to maintain their distance from each other while working. "We only keep one to two personnel up on the Military Kitchen Trailer at once, so we can make sure that they keep their distance," said Dillion Barhorst, army food service specialist. Masking while cooking and serving were also added measures taken to limit the spread of the virus. While not easy, they successfully provided three meals a day to all of the soldiers.

Quoted in Edwin Basa, "Quartermasters Feed Lightning Forge Soldiers Despite Pandemic," DVIDS, July 11, 2020. www.dvidshub.net.

Cooking with Creativity

"My favorite thing that I've made was during deployment. Shrimp on the menu is usually Schezwan shrimp. I decided to change it up a little bit. Our commanding officer is from the South and I'm from the South, so I thought of creating a gumbo, a shrimp gumbo. Rice, red beans, shrimp—all of the above. It wasn't too spicy. It was my first major artwork that I created with a meal. Everybody loved it. Absolutely loved it! Everyone said that was their favorite food they've had on a ship since they've been on here. They've been on the ship for three or four years. They've never had a product that good. It was so good; it made everyone so happy. . . . It was one of my best achievements in the galley."

—Ethan New, culinary specialist seaman assigned to the USS *Preble*

Quoted in Anita Chebahtah, "Culinary Specialists at the Heart of the Crew," DVIDS, September 17, 2018. www.dvidshub.net.

exercises in 2021, his kitchen was in a large tent. In this temporary kitchen, he and the other food service specialists created appetizing meals for hundreds of soldiers in the field, a job they are proud to do. "We are the fuel for the soldiers," says Shastid. "Without food, they won't have the energy to do anything."[23]

Education and Training

Each branch has a specific score required in certain sections on the Armed Services Vocational Aptitude Battery in order to qualify to be a food service specialist. A high school diploma or general equivalency diploma is also required. All enlisted personnel must attend boot camp for whatever service they are in and then attend advanced training. In all branches except the coast guard, personnel attend school for nine weeks at Fort Lee in Virginia. There they learn about nutrition, food preparation, all the basics of cooking, and proper use of equipment. Coast guard specialists attend a fourteen-week course in California. "We have to learn how to shop for food, how to make the menu—all the different components of the meal," says coast guard culinary specialist Katrina Goguen. "You learn everything from soups to sides to entrees. It's a really great foundation for someone who has the interest of one day opening their own restaurant."[24]

Skills and Personality

Cooking is a creative endeavor even when cooking for hundreds or thousands of hungry people each day. "My favorite part about being a CS [culinary specialist], I would have to say, is that I'm able to express myself in a creative way," says navy specialist Ethan New, assigned to the USS *Preble*. "Anytime there's a product that I'm assigned to do, I'm never really basic about it. I like to add a little spice—a little zing zang to the product to make it taste a little better, taste different."[25]

Being calm and focused is also helpful. Kitchens are busy places. Multiple people often work together side by side, sometimes in small spaces, each doing a specific task. Even then, things do not always go as planned. A pot might fall. A loaf of bread might burn. A cook might slice a finger. These things happen in kitchens everywhere. The key to dealing with kitchen crises when they happen is to stay calm and focused.

Knowing how to use time wisely is another essential skill for food service specialists. Navy specialist Roel Cabellero explains:

> Time management is very important in the galley. You have a kettle full of chicken that takes about 45 minutes to cook, but the potatoes have to be seasoned too. So, you look at your staff and see that one-third of your kitchen was sent to the lower-base galley, and you realize you don't have enough hands. Even with all the different tasks I have set before me, the meal must come out right and on time.[26]

Working Conditions
Food service specialists can be assigned to a ship, submarine, or base dining hall or in the field. In all situations, they work in busy kitchens with other food specialists. They use knives, appliances, and other food preparation equipment throughout the day. Those on ships or submarines or in the field must learn to work in smaller spaces even as they make large amounts of food. These specialists interact with the military personnel they are serving and are typically busy from the time their shift starts until it ends. Their tasks range from chopping, boiling, and baking to cleaning counters, washing dishes, and putting away ingredients.

Opportunities for Advancement
Enlisted personnel typically start as an E-1 and may advance as high as E-9. Promotion for enlisted personnel depends on performance evaluation marks, proficiency on exams (which test on career field expertise and knowledge of the military branch), and demonstration of excellent work ethic and performance. As they advance, food service specialists have the opportunity to oversee and guide lower-ranked kitchen personnel and gain supervisory and inspection responsibilities. Titles of higher-ranking jobs include food service division leading petty officer, leading culinary specialist, and shift supervisor.

Employment Prospects in the Civilian World
Working in any kind of food service environment, from restaurants to cafeterias in schools, hospitals, and other institutions, are possibilities after leaving the military. These positions can include food preparation specialists, cooks, sous-chefs, chefs, and kitchen or restaurant managers. Food service experience gained in the military can also lead to specialty jobs such as meat cutter, butcher, or baker.

Finance Officer

What Does a Finance Officer Do?

Finance officers work with money, budgets, and payroll. They oversee purchases of supplies, equipment, and services needed for a military unit's missions. They pay commercial vendors for goods and contractors for services. They balance budgets and make sure purchasing stays within those budgets. Finance officers also are the ones who ensure that operational commands budget for and receive the money needed to run their operations.

While the army uses the term *finance officer*, other branches use different titles for the same job. The air force uses *financial management officer*. In the navy, supply officers also perform finance duties in addition to overseeing logistics.

Air force financial management officer Wendy Kiepke works as the comptroller, a person who oversees the accounting and financial reporting of an organization, for the 104th Fighter Wing in Westfield, Massachusetts. She oversees the finances for the wing and advises the wing's leaders on how to budget for what they need. "I am responsible for the financial resources that come into the unit," says Kiepke. "I manage my office of 10 personnel to ensure they work effectively. Mostly, I oversee the budget to

A Few Facts

Minimum Educational Requirements
Bachelor's degree

Personal Qualities
Detail oriented, focused, logical

Working Conditions
In offices in bases throughout the world

Salary
Depends on pay grade and years of service

Future Job Outlook
Small, competitive branch

Finance Officers Know More than Finance

Military finance officers need to know more than managing finance and budgets. Marine corps finance officers, for instance, also train for the infantry so that they can function in combat situations. Marine corps finance officers assigned to Quantico, Virginia, typically take part in field exercises once a year. During this time, they conduct simulated combat missions, living in the field, using rifles, and taking part in strategic planning. "I think it is important for the finance marines to do this training because is sustains their basic rifle skills and works on the tactics they learned and a lot of times they don't get the opportunity to do that in the office," says Preston Beasley, a marine corps finance officer. They also carry out financial tasks that would be required in a war zone, such as making sure people get paid.

Quoted in Devil Dog, *MCCDC Finance Marines Combat Training in Quantico*, YouTube, April 22, 2016. www.youtube.com/watch?v=pPz83Ry0GwM.

get resources out to the unit level so that our members can execute their missions and the training requirements they have."[27]

Finance officers take care of various financial needs. They are responsible for payroll for their units—making sure that personnel get paid on time and in the correct amounts. On overseas deployments, they might offer help with exchanging US dollars for another country's currency. They also help personnel set up automatic deposits to bank accounts so that their pay goes where they want it to go and in a timely manner.

Typical Workday

On one of her tours in the army, Jennifer Evans was deployed to Camp As Sayliyah in Qatar in 2016. She was commander of the Financial Management Detachment, which provided all sorts of financial services to soldiers. The financial detachment could support up to five thousand service members. It was up to Evans and her team to ensure that all base personnel received their pay in either US dollars or local currency (depending on their stated preference). She worked closely with her staff to make sure they provided accurate advice to soldiers who asked for help. Many service personnel are young and have never had a savings or checking account. So her team helped them understand their options and set up these accounts.

Payments to contractors also flowed through Evans's office. A base with up to five thousand personnel does a lot of business with contractors. For example, contractors provide food supplies and help with some building repairs. "When everyone thinks finance, they just think pay," says Evans. "Really, that's the smallest portion of what we do. When you start looking at paying agents and all the contracts, that's where finance is working in conjunction with units, supporting and working together to get [contracts properly paid]."[28]

Also deployed overseas, air force financial management officer Christopher Wilkes led a financial management office at Al Dhafra Air Base in the United Arab Emirates in 2019. He and his staff oversaw a wide range of financial matters for the leadership and the personnel. His department set the annual budget for the base operations, oversaw the Government Purchase Card (a government credit card that service members use to make purchases for their unit), provided cash transactions to personnel based there, and funded travel of service members there. To keep things running smoothly, Wilkes reviewed the budget daily, assigned tasks to his staff, and reviewed their progress. He describes his work as wide ranging. "Finance touches all base operations—from providing funding for de-rubberizing the airfield to providing cash to Navy SEALs to carry out their mission downfield, finance management is involved in every aspect of the mission!"[29] says Wilkes.

Education and Training

All finance officers must have a bachelor's degree from a college or university and obtain a commission in the military, either through attending a military academy, completing a Reserve Officers' Training Corps program at a college, or completing Officer Candidate School. Finance officers in the army take the Finance Officer Basic Course for thirteen weeks at Fort Jackson in South Carolina. There they learn leadership skills and the financial systems and practices used in a finance platoon. In the air force, financial management officers must have a bachelor's degree in business administration, industrial management, business management, or something similar. Finance officers in the air force and marine corps must also attend a Financial Management Officers Course. For the marines, it is an eighty-five-day Financial Management Officers Course at the Financial Management School at Camp Johnson in North Carolina. For the air force it is for nine and a half weeks at Maxwell Air Base in Montgomery, Alabama.

Skills and Personality

Finance officers must be extremely detail oriented as they track money, pay people, budget, and ensure that units receive the exact amounts requested. They also need to focus on details to understand all of the laws and regulations they must follow regarding finances and government money.

The ability to analyze numbers, budgets, and contracts is another important characteristic for a finance officer. Analysis is key in helping operational units decide how to budget and spend their money. Finance officers look at past expenditures and consider future needs when determining what amounts to allot for equipment, supplies, travel, and other needs.

Finance officers must have good communication skills. They must be able to clearly articulate what they need their team to accomplish and must listen to understand any issues the team has. They deal with customers on base as they assist units with making their budgets, so they must be able to understand the customers' needs while communicating what they can do for the customers.

Finance officers must also be adept at management. As officers, they are typically in charge of others, from enlisted person-

> ### Customer Service Is the Key
>
> "Where we try to shine is customer service. We developed relationships with customers, so we know what they need before they show up. We got a better understanding of their priorities, so we know how to help them."
>
> —Steve Rusnak, air force finance officer
>
> Quoted in Victor Joecks, "Financing the Fight: B/4th FMSD Keeps Funds Flowing," DVIDS, July 21, 2016. www.dvidshub.net.

nel to officers of lower rank. Additionally, they need to be able to motivate their team to complete the needed tasks and work together to accomplish their overall goals.

Working Conditions

Finance officers typically work in offices on bases. Throughout their career, they may be assigned stateside or overseas, depending on where their services are needed. They and their team work on computers for much of the budgeting, payroll, and financial management activities. They also interact with military personnel in need of financial assistance. Additionally, they interact with leaders of different operational departments as they assist them with setting up and following a budget with their assigned department's finances. Generally, their job is during the daytime, although they may have collateral duties, such as physical training or watch officer, that require additional time.

Opportunities for Advancement

Finance officers make up a small community in all branches, so advancement is competitive. Officers typically start at O-1, and the highest level they can attain is O-8. Receiving a promotion to the next rank gets more competitive as they advance. Finance officers need to acquire outstanding fitness reports, engage actively in military life by supporting on-base activities and extra duties, and accept both stateside duties and overseas deployments. As they advance, graduate school is offered to officers, which is necessary to achieve higher-level ranks. Job duties of higher-level officers include comptroller, budget operations manager, defense acquisitions manager, finance battalion executive officer, and finance battalion company commander.

Employment Prospects in the Civilian World

The skills and responsibilities of a finance officer in the military are easily transferrable to civilian careers. Companies looking for accountants, budget analysts, contract supervisors, vendor managers, and other finance-related jobs are open to veterans with financial experience. The government also hires financial managers, analysts, and accountants. Because finance officers have management experience, they are often capable of obtaining high-paying jobs with benefits and a track toward upper-management positions. As an example, Josh Weed served as an active-duty air force financial management officer for ten years. He still serves as a reserve officer, working for the military on weekends or short deployments. Because of his experience in the air force, he was able to obtain a job as a senior director at a company that provides organizational advice to business start-ups.

Public Affairs Officer

What Does a Public Affairs Officer Do?

When the navy decided to begin discharging sailors who would not get the COVID-19 vaccine, it was public affairs officers who released this information to the public through various media outlets. Public affairs officers serve as the mouthpiece for the military. They provide information about military news and actions to the general public, lawmakers, and even other military personnel. These officers work with local media, from newspaper reporters to TV newscasters, to deliver their command's military news. They present information to the public about current military issues and provide insight on how a particular piece of news may affect the military, the community, and the world. "We are the spokespeople who communicate with many different audiences, not just the public—sometimes it's an internal audience, sometimes it's Congress, sometimes it's an international community—but we lead the unit's communication strategy,"[30] explains navy public affairs officer Theresa Carpenter.

Within the military, public affairs officers help prepare higher-level military personnel for news

A Few Facts

Minimum Educational Requirements
Bachelor's degree

Personal Qualities
Personable, good communicator, discerning

Working Conditions
In an office, indoors, stateside or overseas, on an aircraft carrier

Salary
Depends on pay grade and years of service

Future Job Outlook
Approximately 1,340 in the military and growing yearly

media interviews. Generally, it is also the public affairs officers who schedule and arrange press conferences for the senior-level military personnel. Carpenter explains, "We . . . advise the commander on communication policy [and any of] the staff who is going to be communicating with either media or with the public. We are the representatives to help broker those arrangements and to help to help tell our story to say what we're doing say, why we're doing it, and even deal with controversial issues."[31]

Public affairs officers also disseminate information to military members. This might include information about upcoming base activities and classes or new services being offered. It might also involve getting information to personnel about policy changes, health threats, and other concerns. For example, Carpenter had discussions with service personnel about rules and regulations instituted in connection with the COVID-19 pandemic. It was the job of public affairs officers to make sure that all military personnel understood not only the rules but also the reasons for them. "So, if the military makes a policy decision—let's say in regards to Covid—about not going to bars . . . we need to be able to explain to sailors why that's important and in a way that they can understand and that . . . won't cause a lot of controversy,"[32] Carpenter explains. These officers oversee the content and presentation of information in all forms to the military and the public. Public affairs officers ensure people within the military and civilian world understand what the military is doing and why.

A Typical Workday

Army public affairs officer Jonathan Klein was stationed with the 259th Expeditionary Military Intelligence Brigade in Tumwater, Washington, in 2018. It was his daily job to share the brigade's accomplishments with the rest of the army and the public. His duties included observing and reporting on what the brigade did during Exercise Always Engaged. Exercise Always Engaged was a field exercise to prepare the brigade for military intelligence operations, including intelligence collection, analysis, and dissemi-

Creating Podcasts to Share Stories

Podcasts (audio content that can be streamed and downloaded) are popular in the civilian world and now they are gaining traction in the military. Creating and producing military podcasts falls to service members who work in military public affairs. In February 2021 the public affairs office of the US Army Forces Command (FORSCOM) released the first episode of its podcast *FORSCOM Frontline*. The podcast tackles topics that affect army readiness, including racism, extremism, suicide, and sexual harassment and assault. "We want to have real conversations," says Scott Rawlinson, FORSCOM director of public affairs. "We are going to tell the stories and share the experiences of our People—because our Army is comprised of amazing individuals who come together for a common purpose—service to our Nation." As an example, the first episode featured an interview with army general Michael X. Garrett, speaking on the topic of racism and how the army is embracing diversity.

Quoted in FORSCOM Public Affairs, "U.S. Army Forces Command Launches New Podcast," US Army, February 21, 2021. www.army.mil.

---

nation, as if they were in a combat situation. "Their story needs to be told," says Klein, "Nobody gets to see the blood, sweat and tears a unit sheds to make these exercises happen. The long nights, staff exercises, and pain that goes into forming a unit are important experiences."[33] During Exercise Always Engaged, Klein lived the brigade's personnel and observed their exercises as he wrote about the soldiers' experiences.

Like Klein, army public affairs Angelina Cillo works to share her unit's stories. She is assigned to an army group in Cedar Rapids, Iowa. As a reservist, she works a civilian job in Colorado then travels monthly from Colorado to Iowa, where she serves the army for a weekend and sometimes longer. When in Iowa, Cillo's job is to ensure that the group's training is captured and shared with military personnel and the public. To do that, she works with military websites and local news media—writing articles, issuing press releases, and giving interviews. Throughout the day, she supervises enlisted photojournalists and broadcasters and assigns them stories to cover for print, broadcast, and digital media. Sharing the stories allows the US public to understand what the army group Cillo is assigned to is doing in preparation for possible operations overseas. "I love capturing what our Soldiers do while at battle training assembly or annual training," Cillo says. "Soldiers appreciate seeing their photos on social media, they get excited, and it builds esprit de corps within the team."[34] Cillo and Klein both find that sharing the accomplishments of the military boosts the service personnel's confidence and morale.

Education and Training

High school preparation can include taking honors and advanced English courses, journalism courses, and public speaking. Additionally, being involved in a high school newspaper, yearbook committee, or debate club would help develop public affairs skills. Public affairs officers must obtain a bachelor's degree from a college or university. Desirable fields are communications, marketing, journalism, and English. All must receive a commission through a military academy, Reserve Officers' Training Corps, or Officer Candidate School. The branches also require that officers be able to obtain a security clearance due to the information a public affairs officer may have access to.

Officers of all branches complete a twenty-six-week program, the Public Affairs Qualification Course, offered through the Defense Information School at Fort Meade in Maryland. The course covers how and when to release certain types of information, how to implement effective communication strategies using various forms of media, and how communication is integrated with military planning and operations. During the initial part of the course, officers have to pass a writing test because much of the job includes writing press releases, articles, and

A public affairs officer with the New York Army National Guard talks with members of the media. Military personnel who work in public affairs provide information about military events and activities to the general public, lawmakers, and other military personnel.

speeches. Following this course, navy officers attend a ten-day Public Affairs Expeditionary Course focused on how to use public affairs skills in the field.

Skills and Personality

The ability to communicate clearly and concisely, in writing and in speaking, is essential as a public affairs officer. These officers communicate with the military and the public through all means of communication—from newspapers to TV to social media. They also must be able to communicate easily with higher-level officers, giving their advice on how best to present news and information to the public.

Disaster Communication

"One of the most important skills of a PAO [public affairs officer] during a disaster scenario is the ability to establish and sustain a proactive dialogue with the mass media outlets, ensuring the audiences truly understand the role of the Army Reserve forces during the emergency and reinforcing the trust and confidence of the local communities."

—Dustin A. Shultz, army brigadier general

Quoted in US Army Reserves, "Army Reserve Public Affairs in DSCA operations," December 6, 2017. www.usar.army.mil.

As officers, they need to be able to manage their personnel and inspire them to do quality and timely work. "What does it mean to be an officer? It means to serve my people,"[35] says public affairs officer Rachel Buitrago. She explains that this can entail explaining to her public affairs enlistee why they have to come on duty to take photos at three o'clock in the morning so that the enlistee is motivated to do a good job. She believes it is her job to encourage her staff to do the best they can in their jobs and provide them the assistance they need to do so.

A public affairs officer's job also requires creativity in order to produce interesting articles, videos, and digital media. While these need to be accurate and factual, they are also meant to captivate readers and viewers. "Don't be afraid to be creative," Cillo says. "The Army has a lot of parameters but find the angle to your story. Soldiers will appreciate it."[36]

Working Conditions

Public affairs officers can be based in offices both stateside and overseas. They may also be on aircraft carriers, living on the ship, or in the field in a command unit. While they spend part of their time in the office, they also attend many base events, such as changes of commands, athletic competitions, and any event requiring publicity.

Throughout the day, public affairs officers interact with people—whether staff or media or military personnel. Their days are typically busy—filled with events, interviews, information gathering, and writing. During these events, these officers may use or oversee the use of equipment such as still cameras, video cameras, microphones for briefing setups, and audio equipment for podcasts.

Opportunities for Advancement

Officers typically start at the O-1 level and have the opportunity to achieve higher ranks. As they advance, it becomes more competitive to achieve the next rank. To increase their chances, officers should obtain superior fitness reports and serve in different types of duty stations such as in a base public affairs office and as a public affairs officer on an aircraft carrier (if in the navy). Additionally, they should obtain advanced certifications, such as Accreditation in Public Relations from the Public Relations Society of America. Those who obtain higher-level degrees, such as a master's degree, increase their chances for advancement. Types of career opportunities for higher-level public affairs officers include aircraft carrier communications director, chief of media operations, and director of public affairs for a command.

Employment Prospects in the Civilian World

Following military careers, public affairs officers have opportunities in print, digital, and video journalism, public relations, marketing, and communications careers. Overall, the Bureau of Labor Statistics predicts that public relations and public affairs careers will see growth throughout the 2020s, with approximately 7 percent growth for public relations specialists and 9 percent growth for public relations and fund-raising managers.

Source Notes

Introduction: Need for Service
1. Quoted in Christopher Veloicaza, "Logistics Team Keeps 7th Fleet Armed, Fueled, Fed in Indo-Pacific," Commander, US 7th Fleet, July 22, 2020. www.c7f.navy.mil.
2. Quoted in DVIDS, "GW Celebrates 117 Years of Service, Sacrifice," June 18, 2015. www.dvidshub.net.
3. Quoted in Sarah McClanahan, "Only Human—IRT Clinical Psychologist Emphasizes Importance of Mental Health," DVIDS, June 15, 2019. www.dvidshub.net.

Logistics Specialist
4. Quoted in America's Navy, *Navy Logistics Specialist-LS*, YouTube, September 25, 2018. www.youtube.com/watch?v=l6TTYlohNQI.
5. Quoted in Today's Military, "Logistics Specialist: Nathan Grant," 2022. www.todaysmilitary.com.
6. Quoted in Today's Military, "Logistics Specialist."
7. Quoted in Iowa Army National Guard, *92a Automated Logistics Specialist*, YouTube, June 30, 2021. www.youtube.com/watch?v=gx7UKSfRznw.
8. Quoted in Iowa Army National Guard, *92A, Automated Logistical Specialist*, YouTube, August 14, 2020. www.youtube.com/watch?v=8Wocw6NOo8s.
9. Quoted in America's Navy, *Navy Logistics Specialist-LS*.

Nutrition Care Specialist
10. Quoted in Health.mil, "METC Trains Dietician Techs to Build, Support a Medically Ready Force," March 18, 2021. https://health.mil.
11. Quoted in Alexander Singer, "Nutritional Medicine Creates Fit to Fight Airmen," Air Force Medical Service, November 9, 2017. www.airforcemedicine.af.mil.
12. Quoted in Health.mil, "METC Trains Dietician Techs to Build, Support a Medically Ready Force."
13. Quoted in Tegan Kucera, "Nutrition Is a Way of Life," DVIDS, June 11, 2018. www.dvidshub.net.

14. Quoted in Kucera, "Nutrition Is a Way of Life."
15. Quoted in Marcy Sanchez, "LRMC Nutrition Care Expands Capabilities," DVIDS, September 30, 2020. www.dvidshub.net.

Retail Services Specialist

16. Quoted in Chris Jahnke, "Retail Service Specialist a Change for Good," DVIDS, October 7, 2019. www.dvidshub.net.
17. Quoted in Jeremy Beale, "Active Duty Marines Enjoy Work in a Rare Job in the Corps," US Marines, January 11, 2018. www.quantico.marines.mil.
18. Quoted in America's Navy, *Navy Retail Services Specialist*, YouTube, October 8, 2019. www.youtube.com/watch?v=oWR57rXRnW4.
19. Quoted in Jason Pastrick, "Retail Service Specialist Warship 78," DVIDS, October 1, 2019. www.dvidshub.net.

Food Service Specialist

20. Quoted in Matthew Lancaster, "Food Service Specialists Feed the Force," Osan Air Force Base, June 9, 2015. www.osan.af.mil.
21. Quoted in Anita Chebahtah, "Culinary Specialists: The Heart of the Crew," All Hands, September 17, 2108. https://allhands.navy.mil.
22. Quoted in Chebahtah, "Culinary Specialists."
23. Quoted in Caleb Stone, "Food Service Specialists Assigned to the 45th Infantry Brigade Combat Team Ready to Make a Difference," DVIDS, July 18, 2021. www.dvidshub.net.
24. Quoted in David Weydert, "A Culinary Career Worth Serving," US Coast Guard, May 6, 2021. www.mycg.uscg.mil.
25. Quoted in Chebahtah, "Culinary Specialists."
26. Quoted in Travis R, "Navy Culinary Specialists: Career Details," Operation Military Kids, August 13, 2020. www.operationmilitarykids.org.

Finance Officer

27. Quoted in Camille Lineau, "104th Fighter Wing Comptroller Manages Finance, Boosts Morale," DVIDS, March 8, 2020. www.dvidshub.net.
28. Quoted in Victor Joecks, "Financing the Fight: B/4th FMSD Keeps Funds Flowing," DVIDS, July 21, 2016. www.dvidshub.net.
29. Quoted in Mya Crosby, "Finance Management Always Involved with the Mission," DVIDS, March 8, 2019. www.dvidshub.net.

Public Affairs Officer

30. Quoted in Real Sailors Real Sea Stories, *Navy Public Affairs with LCDR Carpenter*, YouTube, October 7, 2020. www.youtube.com/watch?v=hsOpWB3wc_Q.
31. Quoted in Real Sailors Real Sea Stories, *Navy Public Affairs with LCDR Carpenter*.
32. Quoted in Real Sailors Real Sea Stories, *Navy Public Affairs with LCDR Carpenter*.
33. Quoted in John Irish, "PAO Officer Strives to Share Soldiers' Experiences," Northwest Military, October 12, 2018. www.northwestmilitary.com.
34. Quoted in Dino De La Hoya, "The Storyteller's Story: An Afternoon with an Army Reserve Public Affairs Officer," US Army Reserves, August 28, 2020. www.usar.army.mil.
35. Quoted in *CommissionED* (podcast), "059-35P Public Affairs Officer with Capt. Rachel Buitrago," October 21, 2020. www.airforceofficerpodcast.com.
36. Quoted in De La Hoya, "The Storyteller's Story."

Interview with a US Navy Public Affairs Officer

Michaela White is a US Navy public affairs officer. She has served in the US Navy for two years. She answered questions about her career by email.

Q: Why and how did you join the US Navy?
A: I wanted to join the military since I was a young girl growing up with my father, an Army veteran and 1983 graduate of West Point. Watching the camaraderie with his classmates, though they had chosen different life paths, was something that I wanted. I graduated from the United States Naval Academy in 2019 as an Ensign in the US Navy.

Q: How did you train for this career?
A: The Naval Academy [USNA] gave me four years of training ahead of joining the Fleet. USNA has taught me about the bigger picture of the military and helped guide me to transferring into the Public Affairs community in 2020.

Q: How would you describe your typical workday?
A: My typical workday includes going into the office each day and checking in with the fellow officers as well as my enlisted sailors. I serve as a Department Head where I am in charge of 40 sailors who produce various projects such as photos, videos and graphics highlighting the Navy. These projects are shared all over the world. I speak with PAOs across the Fleet to assist with event planning as well as completing qualifications needed as a Junior Officer.

Q: What do you like most about your job?
A: My favorite part about my job is being a Public Affairs Officer in a space that is constantly changing with what is going on in the world as well as the online social media space. This always has me on my toes and adapting to anything that comes my way. I love working together with the other sailors at my command to accomplish missions and ensure that we are showcasing our Navy in the best light. I love the opportunity to teach the story of the Navy through the eyes of its service members.

Q: What do you like least about your job? Why?
A: So far, there is nothing I necessarily like least about my job since I am still relatively new to it. I am working hard to network my relationships to ensure that I am always learning and using that network to answer any questions I may have.

Q: What personal qualities do you find most valuable for this type of work?
A: Personally, the personal qualities that I find most valuable for my specific type of work is being a people person. My job has to do with working with people, both military and civilian, where you must understand that different entities work under different parameters. Going back to being adaptable, you must be able to adapt whatever is thrown your way in order to succeed.

Q: Why do you think the military may be a good option for young adults?
A: I feel that the military is a good option for young adults because it gives you a lot of independence that you may not face in other career fields. You have to be able to adapt yourself and you also get many healthcare and financial benefits with free healthcare, education opportunities, food and housing allowances and job security. Though the job can sometimes be overwhelming, I have a lot more flexibility with my benefits financially than many of my civilian friends.

Q: What advice do you have for students who might be interested in this military career?

A: The advice I would give right now to students who might be interested in this military career is to work hard and stay motivated to achieve your goals. I had my fair share of obstacles in my way to getting where I am now. I am very proud of my perseverance and would tell anyone else that it is totally worth the work to live the life you want, no matter the career path. Stay involved in your community, with your school and build connections all around you.

Other Service Careers in the Military

Air transportation specialist
Automotive mechanic technician
Broadcast journalist
Broadcast specialist
Data processing technician
Dietitian
Disbursing clerk
Finance management technician
Finance specialist
Health services management specialist
Judge advocate general attorney
Legalman
Lithographer
Mass communications specialist
Media communications specialist
Mental health specialist
Motor vehicle operator
Paralegal specialist
Personnel clerk
Pharmacist
Pharmacy specialist
Photojournalist
Postal clerk
Psychological operations officer
Psychological operations specialist
Public affairs specialist
Stenographer
Supply officer
TV production specialist
Warehouse clerk
Yeoman

Editor's note: The online *Occupational Outlook Handbook* of the US Department of Labor's Bureau of Labor Statistics is an excellent source of information on jobs in hundreds of career fields, including many of those listed here. The *Occupational Outlook Handbook* may be accessed online at www.bls.gov/ooh.

Find Out More

Bureau of Labor Statistics
www.bls.gov/ooh/military/military-careers.htm
The Bureau of Labor Statistics is a part of the government that measures labor market activity, working conditions, and salaries, and it provides this info on its website. The military section of its website provides information specific to military careers.

Military.com
www.military.com
This website provides news and resources for military veterans, active-duty military, and those interested in the military. It contains defense news, information on benefits and how to join the military, and more to assist those considering the military.

Today's Military
www.todaysmilitary.com
This is a website produced by the US Department of Defense. On it, one can find out how to join the military branches as an officer or an enlisted person and what types of careers are available.

US Air Force
www.airforce.com
This is the official website of the US Air Force. The careers section provides information on a variety of careers organized by areas of interest.

US Army
www.usarmy.mil
This is the official website of the US Army. The section regarding careers provides information on the different types of careers, how to join, and pay and benefits.

US Coast Guard

www.uscg.mil

This is the official website of the US Coast Guard. In the careers section, one can find information on available careers based on interest.

US Cyber Command

www.cybercom.mil

The US Cyber Command leads the direction of cyberspace operations for the US Department of Defense. Its website has information on how to pursue cyber career opportunities in the different military branches.

US Marine Corps

www.marines.com

This is the official website of the US Marine Corps. It includes a section on roles in the corps, describing the different career paths that are possible.

US Navy

www.navy.com

This is the US Navy's recruiting website. One can find information on available careers, pay and benefits, and requirements to join.

US Space Force

www.spaceforce.mil

This is the official website of the US Space Force. Its careers section contains information regarding both officer and enlisted careers in this branch.

Index

Note: Boldface page numbers indicate illustrations.

Accreditation in Public Relations (Public Relations Society of America), 49
air force
  financial management officers, 37
  Fort Sam Houston, 20
  logistics plans specialist, 9
  service specialists, 30
Allen, Devonta, 29
Armed Services Vocational Aptitude Battery, 20, 34
army, Fort Sam Houston, 20

Barhorst, Dillion, 33
Beasley, Preston, 38
Brandt, Anna, 10–11
Buford, Tawian, 25
Buitrago, Rachel, 48

Cabellero, Roel, 35
Camp Johnson (North Carolina), 40
Carpenter, Theresa, 43, 44
Cillo, Angelina, 45–46, 48
civilian employment prospects
  finance officers, 42
  food service specialists, 7, 36
  logistics specialists, 15
  nutrition care specialists, 22
  public affairs officers, 49
  retail services specialists, 7, 25, 29
COVID-19, 9, 14, 33

Defense Information School (Fort Meade), 46–47
Dwy, Chuck, 4

earnings
  in civilian employment, 7
  finance officers, 37
  food service specialists, 30
  logistics specialists, 8
  nutrition care specialists, 16
  public affairs officers, 43
  retail services specialists, 23
  sample of active-duty personnel, **6**
educational requirements
  finance officers, 37, 39–40
  food service specialists, 30, 34–35
  logistics specialists, 8, 11–12
  nutrition care specialists, 16, 20
  public affairs officers, 43, 46–47

retail services specialists, 23, 26–27
Evans, Jennifer, 38–39
Exercise Always Engaged, 44–45

finance officers
    advancement opportunities, 41–42
    basic facts about, 37
    civilian employment prospects, 42
    educational requirements, 39–40
    job description, 37–39, 41
    personal qualities and skills, 40–41
    titles for, 37
financial management officers, 37
    See also finance officers
Flores, Angelica Gonzalez, 9–10
food service specialists
    advancement opportunities, 36
    basic facts about, 30
    civilian employment prospects, 7, 36
    educational requirements, 34–35
    job description, 4, 30–34, **32**, 36
    personal qualities and skills, 35
    titles for, 30

FORSCOM Frontline (podcast), 45
Fort Irwin (California), 21
Fort Jackson (South Carolina), 40
Fort Lee (Virginia), 12, 34
Fort Meade (Maryland), 46–47
Fort Sam Houston (Texas), 20

Gadson, Christopher, 18
Garrett, Michael X., 45
Gasper, Stephanie, 16, 17
Goguen, Katrina, 35
Grant, Nathan, 8–9, 10

Hampton, Nataliya, 18–19

Ike, Brenda Kedar, 14

Jackson, Linsun, 31
job descriptions
    finance officers, 37–39, 41
    food service specialists, 4, 30–34, **32**, 36
    logistics specialists, 8–11, **12**, 13–14
    nutrition care specialists, 16–20, **19**, 21–22
    public affairs officers, 43–46, 48–49, 53
    retail services specialists, 23–26, **26**, 28, 29
job outlooks
    finance officers, 37
    food service specialists, 30
    logistics specialists, 8

nutrition care specialists, 16
public affairs officers, 43
retail services specialists, 23

Kiepke, Wendy, 37–38
Klein, Jonathan, 44, 45

logistics specialists
  advancement opportunities, 14
  basic facts about, 8
  civilian employment prospects, 15
  educational requirements, 11–12
  job description, 8–11, **12**, 13–14
  personal qualities and skills, 13
  titles for, 9

marines
  community services specialists, 24, 27
  culinary specialists, 30
  finance officers, 38
  logistics/embarkation specialists, 9
Maxwell Air Base (Alabama), 40
McClelland, James, 5
McCray, Racquel, 11
McCray, Tonya, 11
medical services, 5, 14
mental health services, 5

military careers, advantages of, 54
Miller, Patricia, 31–33

Naval Technical Training Center (Meridian, Mississippi), 12, 26
navy
  culinary specialists, 30
  Public Affairs Expeditionary Course, 47
  supply officers, 37
New, Ethan, 34, 35
nutrition care specialists
  advancement opportunities, 22
  basic facts about, 16
  civilian employment prospects, 22
  educational requirements, 20
  job description, 16–20, **19**, 21–22
  personal qualities and skills, 20–21

Occupational Outlook Handbook (US Department of Labor, Bureau of Labor Statistics), 56

Pangborn, Liza, 24
personal qualities and skills
  finance officers, 37, 40–41
  food service specialists, 30
  logistics specialists, 8, 13

nutrition care specialists, 16, 20–21
public affairs officers, 43, 47–48, 54
retail services specialists, 23, 27
podcasts, 45
Polmanteer, Michael L., 21
Profit, Anthony, 17–18
public affairs officers
  advancement opportunities, 49
  basic facts about, 43
  civilian employment prospects, 49
  educational requirements, 46–47
  interview with, 53–55
  job description, 43–46, **47**, 48–49, 53
  personal qualities and skills, 47–48, 54
Public Relations Society of America, 49

Rawlinson, Scott, 45
retail services specialists
  advancement opportunities, 28
  basic facts about, 23
  civilian employment prospects, 7, 25, 29
  educational requirements, 26–27
  job description, 23–26, **26**, 28, 29
  personal qualities and skills, 27
  titles for, 24, 27

Rose, Shane, 18
Rusnak, Steve, 41

service careers
  basic facts about, 4, 5
  types of, 56
Shastid, Darren, 33–34
ship servicemen, 25
  See also retail services specialists
Shultz, Dustin A., 48
Sims, Brandon, 17
Supapo, Joshnino, 24–25
supply officers, 37
  See also finance officers
support services, 5

training. See educational requirements
Trujillo, Evan, 27

US Department of Labor, Bureau of Labor Statistics, 49, 56
USNS *Mercy*, 9–10
USS *America*, 5
USS *Gerald R. Ford*, 11
USS *Harry S. Truman*, 4
USS *Nimitz*, 24
USS *Preble*, 31–33, 35

Weed, Josh, 42
Weed Army Community Hospital Nutrition Care Division (Fort Irwin), 21
Wende, Todd, 5
White, Michaela, 53–55
Wilkes, Christopher, 39

Picture Credits

Cover: Petty Officer 3rd Class Kelsey Culbertson/
       United States Navy

 6: Maury Aaseng
12: MixPix/Alamy Stock Photo
19: AB Forces News Collection/Alamy Stock Photo
26: Operation 2021/Alamy Stock Photo
32: AB Forces News Collection/Alamy Stock Photo
47: Image Vault/Alamy Stock Photo

About the Author

Leanne Currie-McGhee lives in Norfolk, Virginia, with her husband, Keith; children, Grace and Sol; and dog, Delilah. She served as a civil engineer corps officer in the US Navy prior to embarking on a career of writing educational books.